Alan McKirdy has written many popular books and book chapters on geology and related topics and has helped to promote the study of environmental geology in Scotland. His other books with Birlinn include *Set in Stone: The Geology and Landscapes of Scotland* and he is co-author of *Land of Mountain and Flood*, which was nominated for the Saltire Research Book of the Year prize. Before his retirement, he was Head of Know-ledge and Information Management at Scottish Natural Heritage. Alan is now a freelance writer and has given many talks on Scottish geology and landscapes at book festivals and other events across the country.

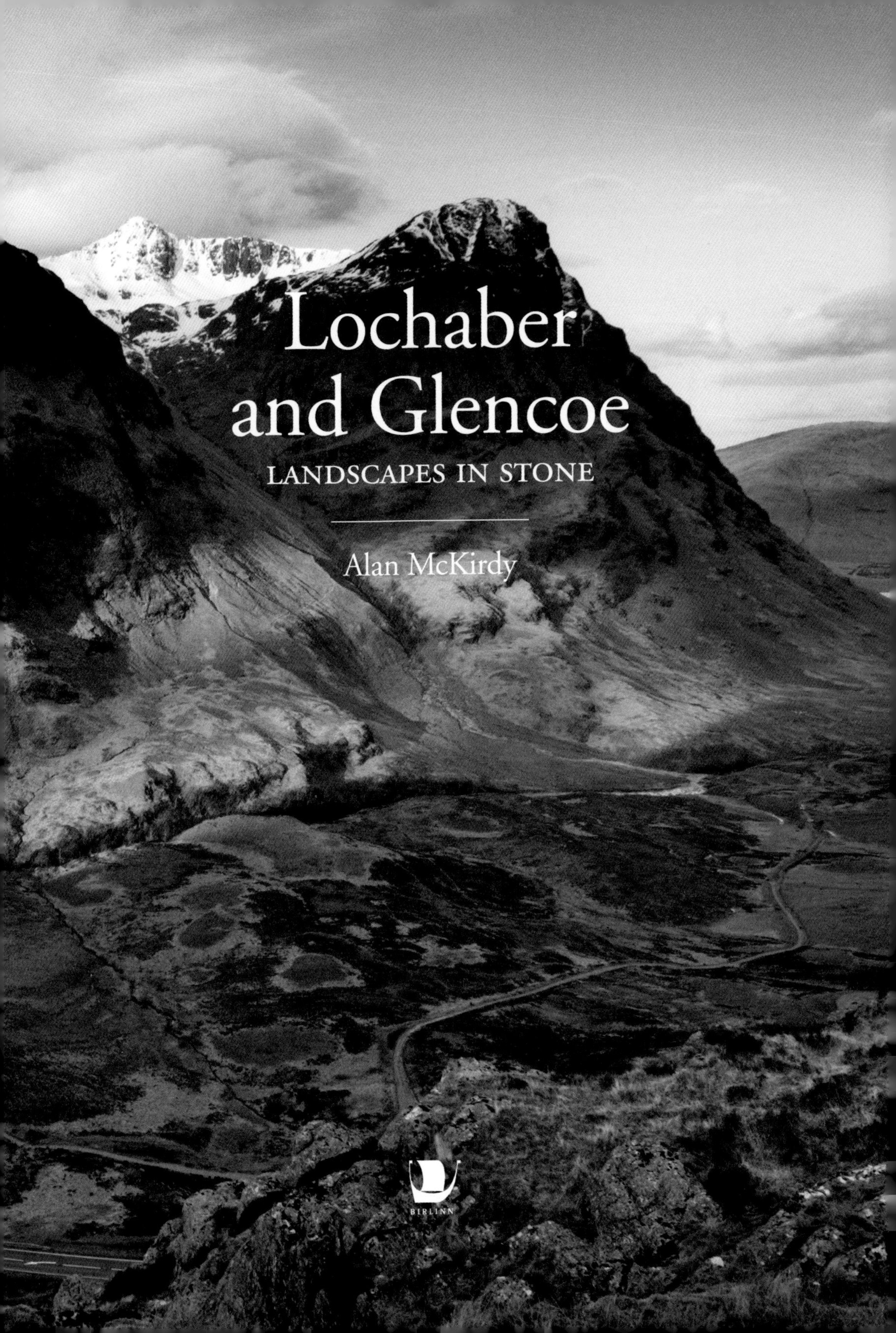

Lochaber and Glencoe

LANDSCAPES IN STONE

Alan McKirdy

BIRLINN

For David Stephenson

First published in Great Britain in 2018 by
Birlinn Ltd
West Newington House
10 Newington Road
Edinburgh
EH9 1QS

www.birlinn.co.uk

ISBN: 978 1 78027 508 6

British Library Cataloguing-in-Publication Data
A catalogue record for this book is available
on request from the British Library

Designed and typeset by Mark Blackadder

Frontispiece: The Three Sisters of Glencoe

Printed and bound in Britain by Latimer Trend, Plymouth

Contents

	Introduction	7
	Lochaber and Glencoe through time	8
	Geological map	10
1.	Time and motion	11
2.	Ancient beginnings	15
3.	Volcanoes rocked that world	23
4.	A world first at Strontian	29
5.	Mind the gap!	30
6.	Age of ice	31
7.	The landscape today	35
8.	Places to visit	42
	Acknowledgements and picture credits	48

Introduction

Opposite.
Buachaille Etive Mòr.

The Lochaber and Glencoe area is rich in historic places, myths and legends. More than three centuries ago, in an act of barbarism, an attempt was made to subjugate the free spirit of the MacDonalds of Glencoe, when they were set upon in their beds and murdered by agents of the British crown. And later, it was at Glenfinnan that Bonnie Prince Charlie raised his standard and started his long march southwards to prosecute his claim to the British throne. This marked the start of the 1745 Rebellion that shook the establishment to the core.

But Lochaber has a longer and even more turbulent history that is played out in the record etched into the rocks. Some of the oldest rocks in Scotland built the western part of Lochaber, buckled, folded and tormented by later movements in the Earth's crust. Ancient volcanoes erupted here in explosive fashion; their shock waves would have reverberated around the planet. We see their eroded remains in Glencoe and also in the highest mountain of Scotland – Ben Nevis. More recent volcanic events are represented in the south of Lochaber where lavas erupted from the Mull volcano 60 million years ago and flowed across the ancient landscape.

'Scotland' was once five separate chunks of the Earth's crust that came together some 420 million years ago. When these fragments of crust joined together, their separate parts (called terranes) had different geological pasts, and the boundaries between them were fault-lines. We now recognise one of the boundaries that separated these ancient terranes as the Great Glen. This deep fracture slices across the area from the south-west to the north-east and is one of Scotland's most prominent natural landscape features. This weakness in the Earth's crust was later excavated further as glaciers scraped their passage across the landscape, carving deep glens and shaving the tops off even the highest of our mountains.

This book unravels some of these geological mysteries and presents a coherent story of how the landscapes of Lochaber and Glencoe came to look the way they do today.

Lochaber and Glencoe through time

Period of geological time	*Millions of years ago*	*Scotland's global position*	*Environments and events in Lochaber and Glencoe*
Anthropocene	Last 10,000 years	57° N	During this time, *Homo sapiens* (people) appeared in this area and started to modify the landscape by clearing the forests and growing crops.
Quaternary	Started 2 million years ago	Present position of 57° N	• **12,500 years ago** – this time was marked by a final advance of the ice, as the climate cooled. • **14,700 years ago** – the climate at this time was similar to that of today; the glaciers had diminished in extent. • **22,000 years ago** – a thick sheet of ice covered Scotland that extended 100km beyond the present-day coastline. • **Before 29,000 years ago** – there were many advances and retreats of the ice, separated by warmer interludes, known as interglacial periods.
Neogene	2–24	55° N	Subtropical conditions prevailed during this time. At the end of this period, the climate cooled as the Ice Age approached.
Palaeogene	24–65	50° N	Volcanoes erupted in adjacent areas, such as Mull and Skye. Some lavas flowed into the land that became Lochaber.
Cretaceous	65–142	40° N	Warm shallow seas covered most of Scotland. Limited deposits of this age are preserved around Lochaline and Beinn Iadain.
Jurassic	142–205	35° N	Dinosaurs roamed the place we now recognise as the Isle of Skye but no evidence of rocks of this age is preserved in the Lochaber and Glencoe area.
Triassic	205–248	30° N	Seasonal rivers flowed across the area.

Period of geological time	*Millions of years ago*	*Scotland's global position*	*Environments and events in Lochaber and Glencoe*
Permian	248–290	20° N	Small patches of sediments from these times are preserved in the south of Lochaber.
Carboniferous	290–354	On the Equator	Scotland lay close to the Equator at this time and tropical rainforests were widespread across the country.
Devonian	354–417	10° S	The Old Red Sandstone continent was created after continents collided.
Silurian	417–443	15° S	The collision of continents was complete in Silurian times. It involved the coming together of five separate chunks of land called terranes. Violent explosions occurred as Glencoe and Ben Nevis formed as volcanoes.
Ordovician	443–495	20° S	Colliding continents started to form a mountain chain of Himalayan proportions.
Cambrian	495–545	30° S	The Iapetus Ocean started to close.
Proterozoic	545–2,500	Close to South Pole	Sand, mud and limestone accumulated on the edge of a continent known as Laurentia. This pile of sediments was later cooked and squashed as continents collided. Between 1,000 and 870 million years ago the Moines were deposited as sand and mud.
Archaean	Prior to 2,500	Unknown	Small patches of Lewisian gneiss were present alongside the Moines of the Morar area.

The geological map of Lochaber and Glencoe is an elegant construct. It represents some of the most ancient rocks in Scotland. From west to east, tiny patches of ancient Lewisian basement rock (gneiss) sit juxtaposed with the Moine schists of slightly younger age – see later explanation. To the east of the sharp break of the Great Glen Fault, now occupied by the waters of Lochs Linnhe and Ness, the Dalradian dominates. The ancient rocks of the Dalradian were deposited on the edge of a continent known as Laurentia and were then folded as continents collided some 420 million years ago. Through this contorted basement rock, the Ben Nevis and Glencoe volcanoes punched their way to the surface creating seismic waves aplenty in the process. Tiny patches of sedimentary rocks from the Carboniferous, Permo-Triassic and Cretaceous Periods are present, providing tantalising glimpses of a wider geographic coverage of strata (layers of rock) before they were pared back by erosion. The Mull volcano erupted around 60 million years ago, covering the southern part of the area described in this book with molten lava that flowed northwards across the ancient landscape. The area was covered by ice for much of the last 2.6 million years. Valleys and mountains (glens and bens) were created by ice sheets and rivers of ice that moulded the countryside into the familiar contours of today's landscape.

1
Time and motion

Time

One of the most challenging aspects of studying geology is gaining an appreciation of the timescales involved. Geologists refer to events stretching back millions or even billions of years as if they were commonplace, but timescales like this don't feature much in our everyday conversations. We tend to see historical events as measured in human timescales, and events that took place before recorded history, such as the building of the Standing Stones of Callanish on Lewis or the beautifully preserved houses of Skara Brae on Orkney, are about as far back as we can comfortably imagine.

To understand the geological development of this place – the Lochaber and Glencoe area – we must be prepared to grapple with

Ben Nevis is pre-eminent amongst Scottish mountains. At 1,345m (4,411 feet), it towers above the rest. Magnificent views such as this are sometimes described as 'timeless', but rocks that built our landscapes can be dated accurately, and the sequence of events that gave rise to this landscape is better understood as a result.

some pretty big numbers. The Earth was formed some 4.5 billion (thousand million) years ago. Thin slivers of some of the very earliest rocks to form part of the planet's early crust are preserved near Mallaig. A conservative date for these rocks is around 2 billion years old, so we travel back to a time when Planet Earth was an inhospitable place, completely unrecognisable from the land we call Scotland today.

Dating rocks and associated geological events is a relatively recent innovation. This technique allows geologists to calculate the age of rocks with a reasonable degree of accuracy: within a few million years is a good result! Geochronology, or the accurate dating of rocks, was pioneered in the 1950s by Professor Arthur Holmes, Regius Professor at the University of Edinburgh. His idea was a simple one. Many rocks contain minerals that are naturally radioactive. Geologists can isolate a crystal of one of these radioactive minerals, uranium for example, which decays to lead over a known time period. By measuring the quantities of uranium and lead in a sample, the time over which that transition from one element and to another has been taking place can be calculated. Add in some basic maths, and the date of formation for that mineral is then known. This technique opened up a whole new way of understanding the rocks that make up our world and allowed us to unravel and understand the geological events that shaped the landscape of this area.

It also allowed us to divide up geological time into coherent chunks as described in the table presented on pages 8 and 9. Each geological period (Cambrian, Ordovician, etc.) is time-bounded by internationally agreed numbers that define the start and end of each slice of geological time. Arthur Holmes' preoccupation with geological time has helped to unlock many secrets about the way our world developed.

Charles Darwin was the scientific doyen of his age. His ground-breaking book *On the Origin of Species* was ridiculed by some, but eventually became one of the best-sellers of the age. Darwin visited the Lochaber and Glencoe area on another scientific quest to consider the way in which the Parallel Roads of Glen Roy were formed (see pages 33 and 34).

This extended timescale also allowed plants and animals to evolve from the simplest single-cell organisms a billion years ago to the dazzling complexity of biodiversity that we see around us today. Charles Darwin and his less celebrated contemporary Alfred Russel Wallace, both working in the mid nineteenth century, were the first to realise the possibilities that 'evolution by natural selection' had to offer. Key to this theory, perhaps the most important scientific idea of that century, was a requirement for individual species to develop, change and therefore evolve over successive generations. This required time – lots of it. Through the work of earlier Scottish scientists, notably James Hutton and Charles Lyell, Darwin understood that the geological record stretches back through aeons of geological time, thus comfortably accommodating the radical new idea of evolution. The precision that Arthur Holmes' development of geochronology offered a century later was a further confirmation of these early pioneers' work.

Motion

Turbulence in layers of the planet beneath our feet has propelled Scotland and every other slab of the Earth's crust around the surface of the globe since the very earliest of times. The outer skin of the planet, known as the crust, is divided into large sections known as tectonic

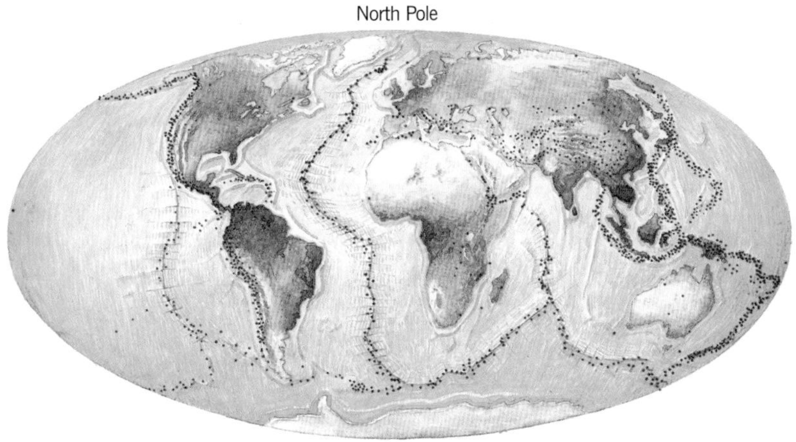

The edges of the Earth's tectonic plates are defined by the red dots on this world map. Each dot represents places where earthquakes frequently take place. The shock waves created by earthquakes are known as seismic activity.

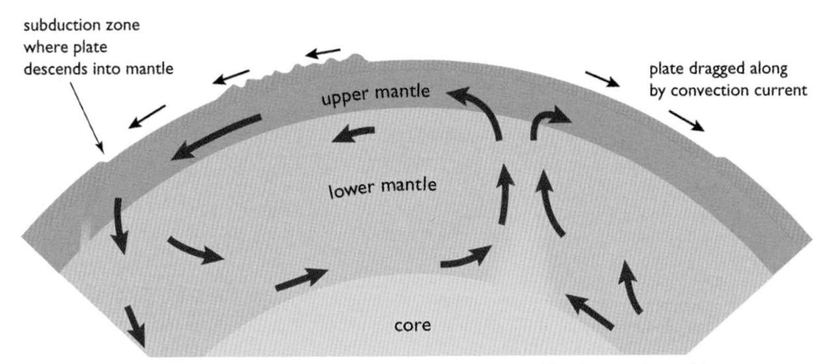

Heat from the Earth's core drives the motion of the continents. The temperature of the core is estimated to be around 6,000°C. This heat flows outwards to the mantle, the layer between the core and crust, and sets up a convective motion that moves the continents along at a rate of around 6cm per year. Over many millions of years, this driving force has created mountains, caused countless natural disasters and moved Scotland from the South Pole through every climatic zone to its current location at 57° north of the Equator.

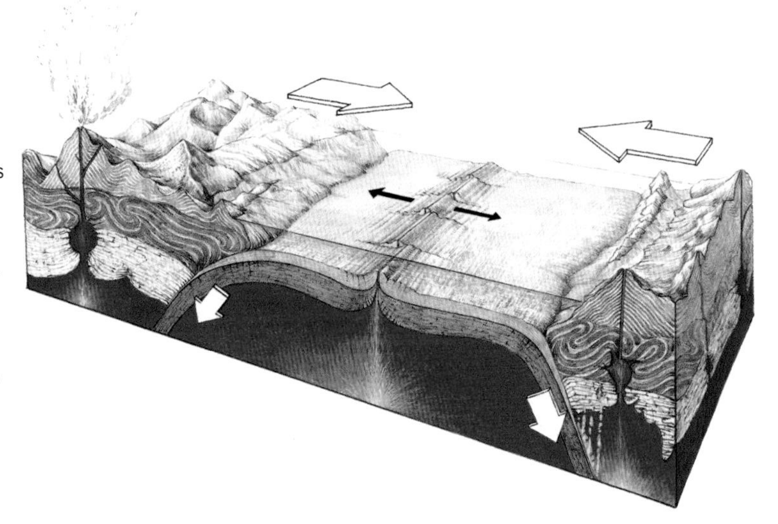

Boundaries between the Earth's plates are extremely active. Driven by the heat from the Earth's core, the plates bump and grind past each other, creating earthquakes or seismic activity in the process. In places, crust is destroyed as one plate dives under its neighbour. This is balanced by the creation of new crust, mainly at the mid-point of the world's oceans, where a string of volcanoes spew new volcanic rock onto the sea floor and, in so doing, drive the continents further apart.

plates. There are seven large ones and around a dozen smaller plates that make up the crust.

The result of this dynamic activity is that continents move huge distances over protracted periods of geological time, oceans come and go like seasonal puddles and new mountain ranges are created. As continents converge, the layers of sand, mud, limestone and lava that have accumulated on the sea floor over many millions of years are caught in a vice-like grip. They are folded and buckled as the continents collide and new mountain chains are formed. The Himalayas, Urals, Alps and Pyrenees were all formed in this way, powered by movements in the mantle. Head-on collisions between colliding continents were inevitable, and the globe is littered with examples of where this has happened in the geological past. As we shall see in the next chapter, the Highlands of Scotland were formed in a similar manner.

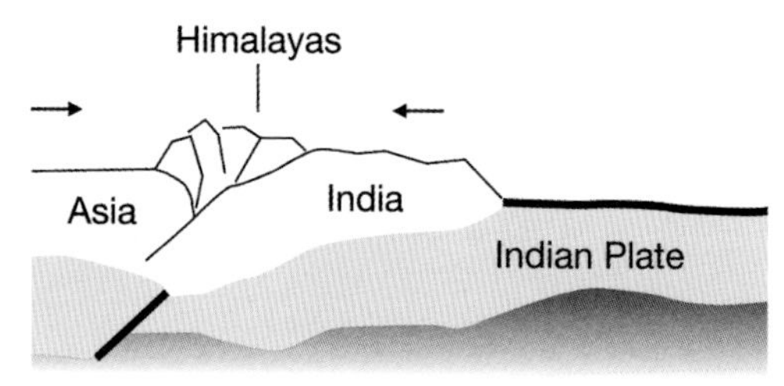

India was driven northwards, colliding with Asia around 55 million years ago. The sediments that collected in the intervening, and now long disappeared, ocean were buckled, folded and thrust upwards to form the world's highest and most imposing mountain chain: the Himalayas. This is a good parallel with how the Scottish Highlands were formed around 420 million years ago.

2
Ancient beginnings

As can be seen from the geological map, the Great Glen slices the Lochaber and Glencoe area in two. On the north-west side, the land is largely built from the Moine, and on the south-east side, Dalradian rocks dominate. Both rock sequences have been cooked and squashed in the white heat of continent-to-continent collisions with the resultant rocks significantly changed or, to use the technical term, metamorphosed. Both Moine and Dalradian rocks were originally laid down as layers of sediment – predominantly sands, muds, limestone with some lavas. Both have been altered by deep burial in the Earth's crust. This caused them to lose much of their original nature only to take on the more banded appearance that is characteristic of a rock type known as a schist. Both Moine and Dalradian rocks are hard and resistant to

This is how the Highlands would have looked after the mountains that underlie the Moine and Dalradian were formed.

erosion as a result of their ordeal by heat and burial. The mountains formed as a result of the metamorphic process would at one time have been as high as the Alps or Himalayas and were only reduced in size and scale by subsequent erosion by wind, water and ice.

A quick guide to metamorphism

Ancient layers of sediment are fundamentally altered during the process of metamorphism. All rocks are composed of minerals as their basic building blocks and that is how one rock is distinguished from another. As continents collide, layers that were deposited on the sea floor become buried deep within the Earth's crust. They are also subjected to an increased temperature. So the application of hugely increased temperatures and pressures progressively changes the mineral composition of the original sediments in response to the new conditions. Deeper burial gives rise to a higher degree of alteration, as the temperatures and pressures become greater with depth. The mineral composition of the rock changes in a manner that is characteristic of a sliding scale of alteration from moderate to severe. The appearance of typical minerals that are found in metamorphic rocks, such as garnet, sillimanite and kyanite, can be replicated in the laboratory, so the temperature and pressure regimes that created them can be accurately modelled. This allows a reliable assessment to be made of the depth to which the sediments were buried in the Earth's crust, which helps geologists to reconstruct the sequence of events that gave rise to rocks we see today.

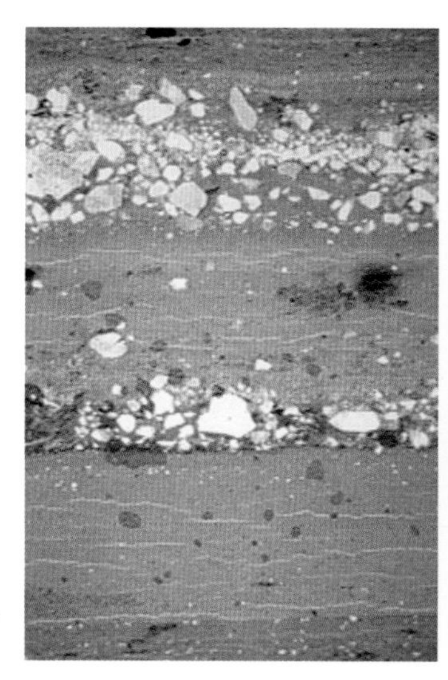

Thin slices through rocks, examined under the polarising microscope, are used to identify the mineral components of specimens that otherwise look dull in lumps of rock the size of your fist. On the left are layers of sand, mud and pebbles (around a metre in thickness) that are similar to the parent material that became the Moines. After the rocks have been altered by heat and pressure at depth in the Earth's crust, the chemical elements of the original minerals are reconstituted into a new mineral assemblage found in that more extreme environment. Through the application of heat and pressure, these layers of sand, mud and pebbles are transformed into the metamorphic rock known as a mica schist (right-hand picture), characteristic of the Moines – as viewed under the microscope.

Another tell-tale sign of the fact that a rock has been metamorphosed is that it presents as buckled, banded and folded. The degree to which this rock must have been reduced to the state of a lump of pliable putty is clearly evident. Just imagine the temperature to which this rock must have been exposed to achieve this effect. This folding is on a small scale, but landscape-scale folds with amplitudes of many kilometres were also formed in this manner. This image shows a section of rock that is around a metre across.

The Moines

Even after a century of study, the geological history of the Moine schists has not been fully unravelled. There are still matters on which experts hold differing views, but the areas of consensus greatly outweigh those where uncertainties have still to be resolved. What we know 'for certain' allows a coherent story to be told about the origins and subsequent history of these rocks. Part of the problem lies in the fact that the Moines have been involved in many metamorphic episodes, so evidence from the previous event is frequently 'overprinted' by the next. Also the Moines have no surviving fossils, or more likely had none in the

The hills that surround Loch Arkaig, north of Fort William, are built from Moine schists and related rocks.

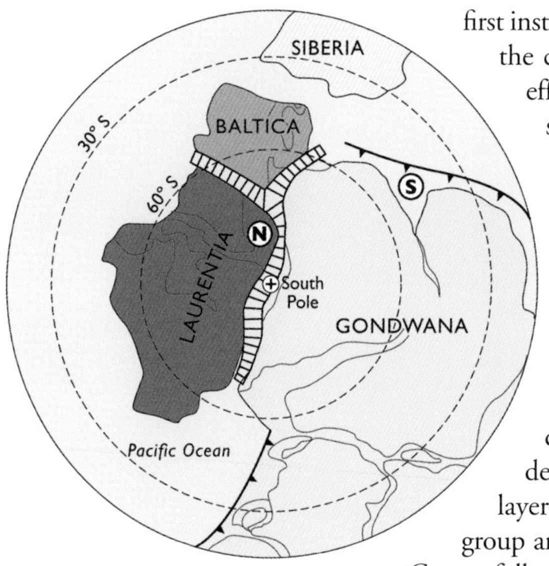

Above. The world was a very different place when the Moine rocks were laid down under a sea as a series of sands and muds. The current consensus among scientists studying these rocks is that the sea in which the Moine layers of sand and mud were deposited was located close to the junction between the major continental blocks of Laurentia, Baltica and Gondwanaland.

These continents were collectively known as Rodinia and had coalesced around the South Pole. The N and S on this map show the relative positions of northern and southern Britain at the time.

first instance. Study of fossil assemblages is the commonest and one of the most effective methods of relating a rock section in one area to another of similar age many kilometres away.

The sedimentary layers that gave rise to the Moine schists were laid down on the sea bed 1,000 million to 870 million years ago.

Although the Moine has been much changed since it was deposited as sands and mud, a definite ordering in the original layering can be detected. The oldest group are sediments known as the Morar Group, followed by the Glenfinnan and finally the Loch Eil Group. Estimates as to the original thickness of sediment that accumulated on the sea floor vary, but a wedge of sands and muds of around 10km is probably about right.

Over the last 1,000 million years ago, the history of these rocks has been eventful. Around 870 million years ago, base-rich pulses of magma (volcanic rocks rich in calcium and magnesium) were introduced, as were ancient granites. Later, the piles of Moine sediments and igneous rocks were metamorphosed, which involved intense folding, addition of more granite and associated faulting.

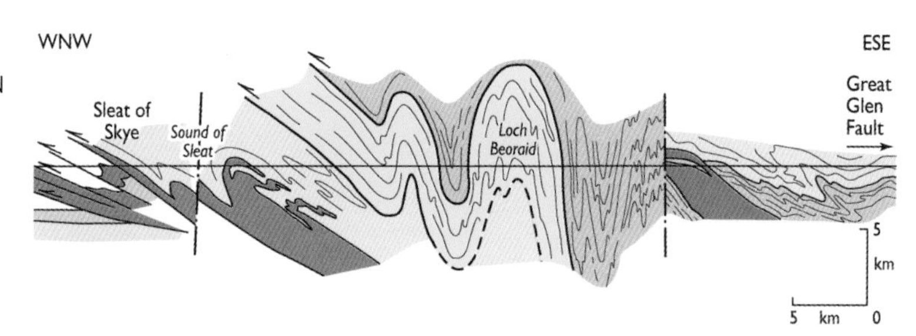

The Moine's torrid history of deep burial, folding, faulting and introduction of molten rocks led to the emergence of a very complex picture. The rocks are folded on both a regional and smaller scale as a result of these metamorphic events. This cross-section through the area from the Great Glen westwards gives an impression of the complexity involved. The various colours identify the component rock units, including the Morar, Glenfinnan and Loch Eil groups.

The Dalradian

Around 600 million years ago, the ancient super-continent of Rodinia broke up and a new ocean came into being – the Iapetus Ocean. It was in a sea that lay marginal to this great ocean that the rocks to the south of the Great Glen were deposited. They are known as the Dalradian. These strata are younger in age than the Moines, but have a similar history in that they were laid down as layers of sands, muds and limestones at the bottom of a sea, later to be cooked and squashed in the bowels of the Earth. Today, the Dalradian occupies the land that lies between the Great Glen and the edge of the Highlands as defined by the Highland Boundary Fault.

The Iapetus Ocean existed for around 250 million years and, during that time, great quantities of loose sediments built up on the sea floor.

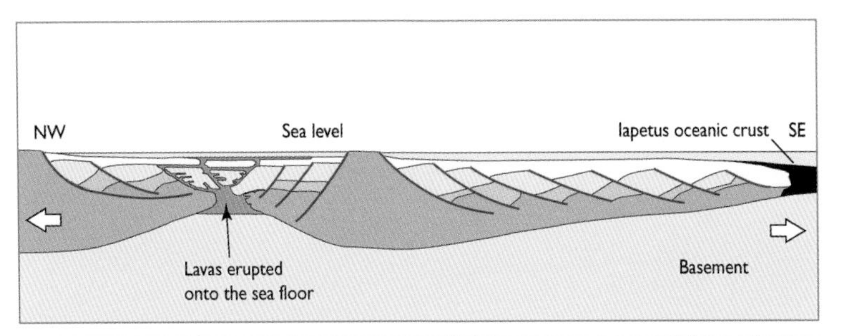

An area adjacent to the Iapetus Ocean (which opened later) was the place where the sediments that became the Dalradian were laid down. Sands and mud built up on this new sea floor, as rivers transported their burden of sediments carried from the upland areas of Rodinia to the sea. These layers were metamorphosed many millions of years later when this seaway closed as continents collided.

Landscapes near Ballachulish, including the disused slate quarries and Glencoe beyond, are carved from Dalradian rocks.

The Iapetus Ocean narrowed and the continents collided with each other around 420 million years ago. It was a slow-motion 'car-crash' that led to the formation of a mountain range the size of the present-day Himalayas. An island arc developed above the subuction zone and was first to collide with Laurentia.

The ocean reached its maximum extent during Ordovician times. The force that drove continents apart went into reverse as the plate beneath the ocean was consumed in a subduction zone, where one colliding plate bends and slips under its neighbour, sinking into the mantle. This sounds complicated, but the diagram below illustrates what happened. This had the effect of bringing the continents on either side of the ocean ever closer to each other and squashing the sediments on the sea floor as if they were being compressed in a vice.

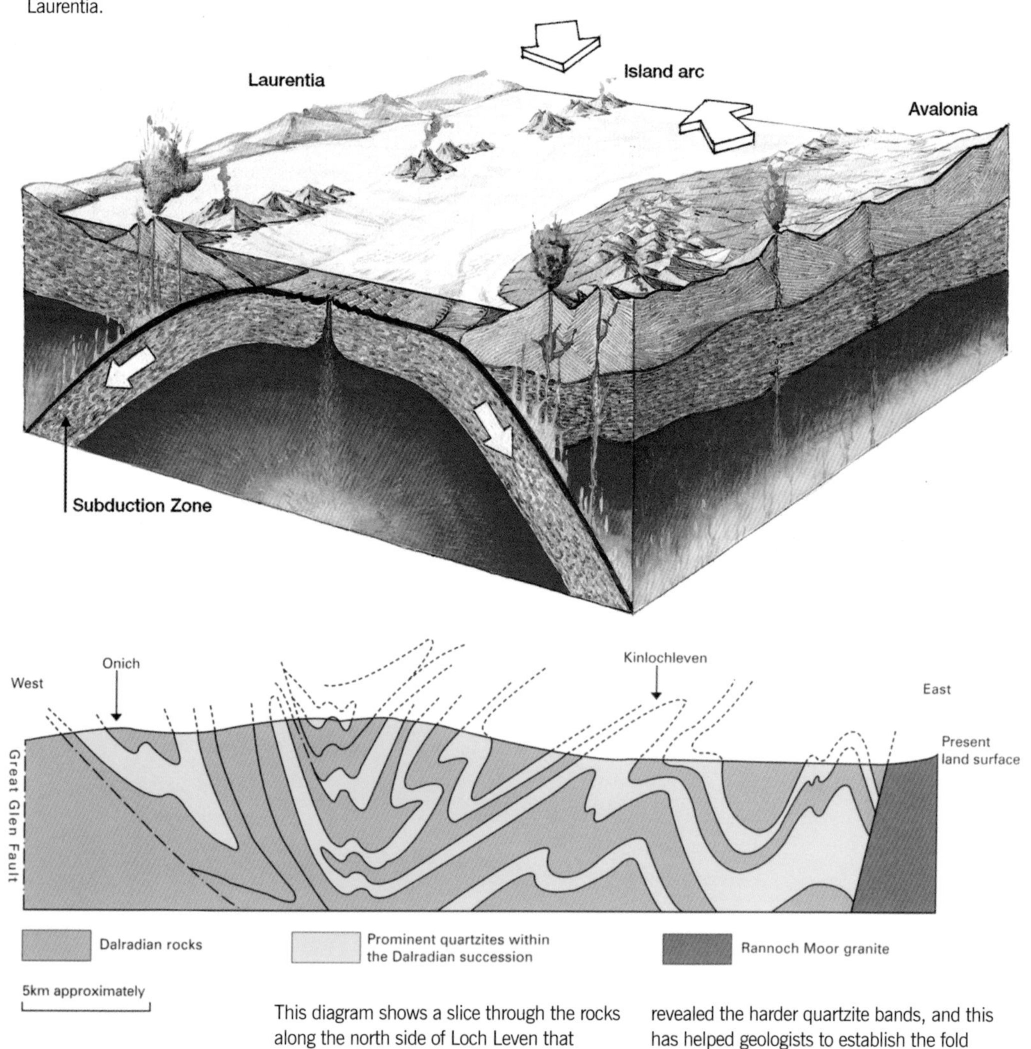

This diagram shows a slice through the rocks along the north side of Loch Leven that demonstrates the large-scale folding that has taken place. Erosion at the surface has revealed the harder quartzite bands, and this has helped geologists to establish the fold structures that the continental collision has created.

By this process of metamorphism, muds were changed into slates and schists, and sands were transformed into quartzites. The Mamore range, which runs parallel to Loch Leven and separates Ben Nevis from Glencoe, consists of a series of quartzite peaks, including Stob Bàn (Gaelic for 'white peak') and Binnein Mòr, which soars to a majestic 1,130 metres.

The quartzite peaks of the Mamores are a spectacular sight.

Assembly of Scotland

The closure of the Iapetus Ocean took place over a protracted period – around 70 million years. It was not a simple affair. Laurentia and Avalonia, which hosted Scotland and England respectively, approached each other in an oblique fashion. Recent research has established the remarkable fact that Scotland was indeed made from five separate continental fragments, known as terranes. Each terrane is separated from the next by a major fault – a terrane boundary. The process worked on the principle of plate tectonics, but in miniature.

The Pap of Glencoe and the adjacent peak of Sgorr nam Fiannaidh are also built from quartzite. This rock is hard as flint and has resisted subsequent erosion by ice, wind and water to remain prominent features of the landscape.

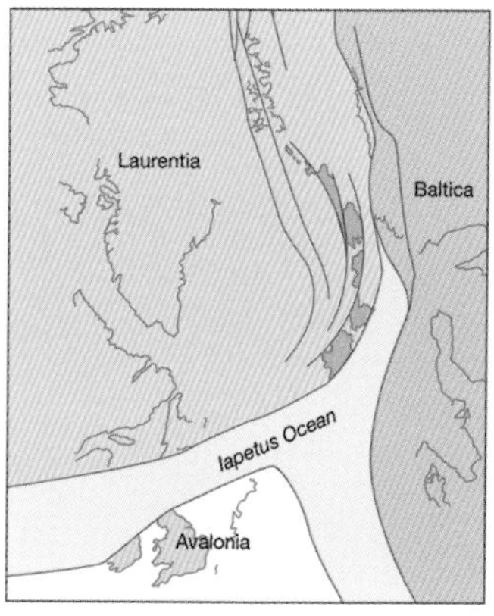

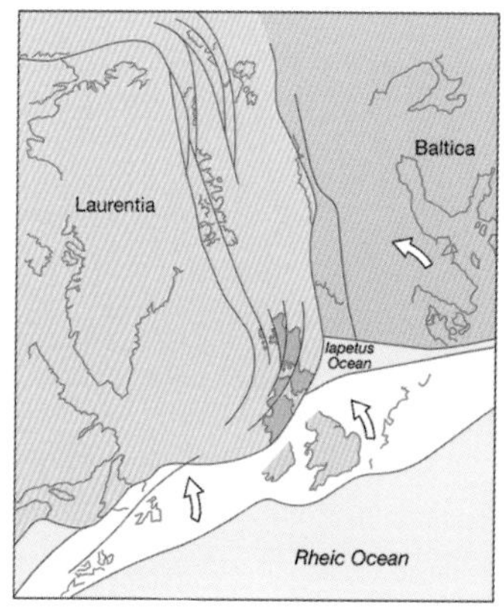

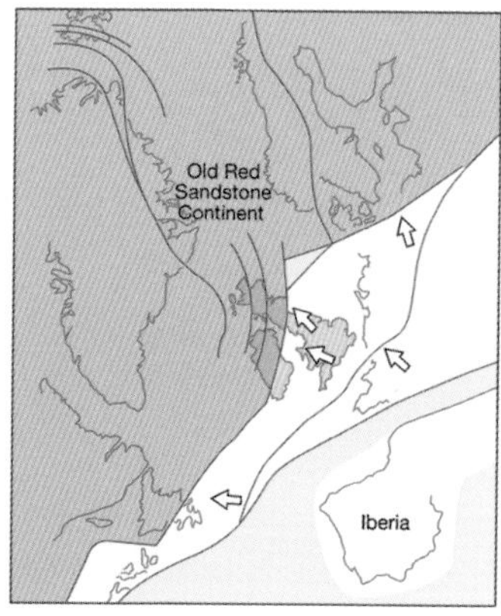

As the Iapetus Ocean closed, the component parts of the land that would become Scotland slid into position. The terranes were all bounded by the major faults that defined each of these micro-continents. These events were significant in the Lochaber and Glencoe area, as one of those cracks that reached deep into the Earth's crust we now recognise as the Great Glen Fault. It slices Lochaber in two, bringing together Moine and Dalradian bedrock. Evidence of movement along the Great Glen Fault is provided by the existence of a rock known as 'mylonite' that is composed of crushed and milled rock fragments. This crushing and fracturing would have happened as the continental plates ground past each other before the complex system of fractures finally locked into place around 420 million years ago. This final chapter, which saw the death of an ocean, also heralded a period of relative stability when a new continent came into being.

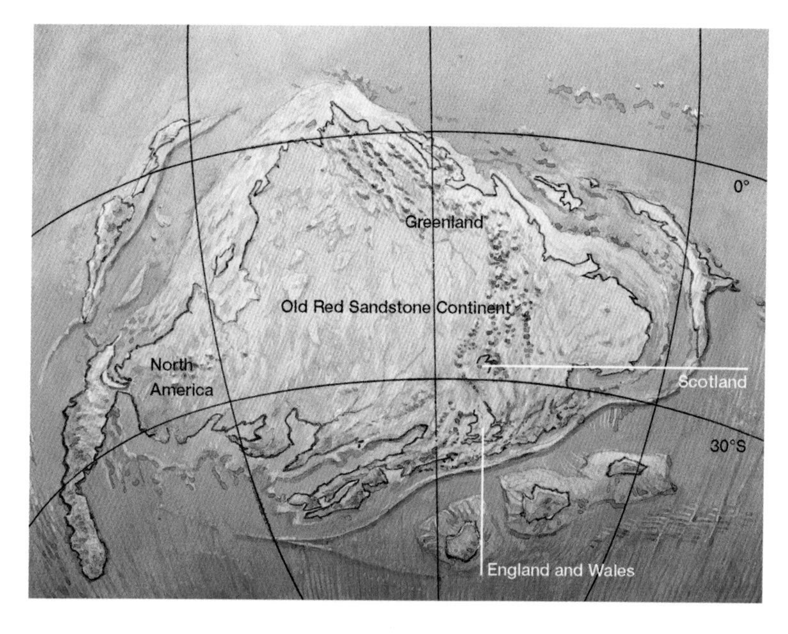

The Old Red Sandstone continent was a product of a global re-arrangement of world geography. Continents collided, an ocean disappeared and the resultant landmass held Scotland close to its heart. This landmass was still located south of the Equator, although its trajectory was ever-northwards.

3
Volcanoes rocked that world

As a result of the white heat of continent-to-continent collision, great volumes of molten rock or magma were produced. A huge number of granites and related rocks pockmark the Scottish landscape today, and the origins of many can be traced back to these apocalyptic events.

Two of these ancient granites are of particular interest – Glencoe and Ben Nevis. Both were active volcanoes from 425 million years ago onwards and their eruptions were particularly explosive. It is no exaggeration to say that, when they blew their tops, both would have created seismic shockwaves that would have travelled around the planet.

Glencoe is a tranquil place today, popular with tourists visiting one of Scotland's best-loved destinations. But some 420 million years ago,

The lower part of the west face of Aonach Dubh shows the molten rock that didn't quite make it to the surface, interspersed with sedimentary layers laid down by rivers. The upper part of the cliff is made from deposits carried by pyroclastic flows that erupted as billowing clouds of searing hot material that emanated from the volcano.

it was a different story. The initial rumblings of the Glencoe volcano were low key. Lava ascended from the lower reaches of the Earth's crust to a level that was close to the surface. At this point, there were no explosive eruptions onto the ancient landscape as it existed at that time. Instead, pulses of molten magma were shot sideways beneath the surface, creating horizontal sheets of rock when they cooled. The sheets of igneous rocks were thrust between sedimentary layers that had been laid down by rivers that criss-crossed the Old Red Sandstone continent. These river deposits consisted of boulders and smaller rock fragments that had been transported to lower ground during flash-flooding events.

The upper part of the cliff face of Aonach Dubh, on the south side of Glencoe, represents a much more violent episode in the development

This super-heated pyroclastic flow, consisting of ash, boulders and steam, erupted and travelled down the slope at Mount St Helens, Washington, at terrifying speed. There were not many casualties associated with this eruption in 1980, but those that did occur were because of the speed at which the gas cloud moved. It engulfed those who had not managed to move to a safe distance from the volcano.

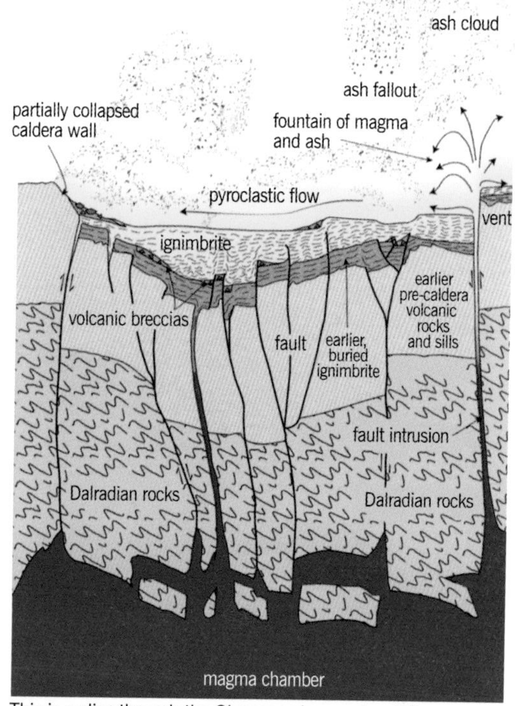

This is a slice through the Glencoe volcano when things were at their most dramatic. As the magma chamber emptied its contents of molten rock, the roof collapsed to fill the void. The surface expression of this collapse was a crater, known as a caldera, filled with boiling rock, ash and steam. The caldera would have been circular or oval in plan, with faults defining its outer limit. These faults created edges that were steep-sided and inherently unstable, so large blocks of rock would have fallen into the boiling cauldron below.

of the volcano. Pyroclastic flows, similar to those that buried the towns of Pompeii and Herculaneum almost two thousand years ago, scarified and scorched the early landscape of this area. The super-heated clouds of gas, volcanic ash and rock fragments now form a layer of solid rock that is in excess of 150m in thickness.

This exposure of rock face on the slopes of Buachille Etive Beag shows the angular fragments of rock and volcanic ash that were carried along within the pyroclastic gas cloud.

The overall scale of the cone-shaped structure (edifice) of the volcano was massive – estimated to be around 14 km in length by 8 km wide. The volcanic activity lasted for tens of thousands of years. But each individual episode would have been relatively short-lived. Rivers continued to run across this volcanic landscape, and large lakes formed within the caldera itself. The presence of so much water made the volcanic episodes even more explosive. As the hot sticky magma rising from the depths came into contact with the saturated ground above, water was converted into super-heated steam with a resulting huge expansion in volume. This smashed the magma apart, producing huge volumes of ash and a greatly increased likelihood of creating further pyroclastic flows.

Above. Some of the lavas cooled slowly to create these beautiful columns, up to 200m high, as seen on Stob Coire nan Lochan, Glencoe.

Ben Nevis is Scotland's highest peak. It has been assailed by the erosive forces of ice, wind and water since it erupted over 425 million years ago, but it still soars to an unequalled 4,411 feet or 1,345 metres above sea level.

The Ben Nevis volcano was also a caldera where a central block subsided into the magma chamber below. There are two generations of granite that can be recognised today and they were emplaced thousands, perhaps even millions, of years apart.

Right. Volcanic rocks form the north-eastern buttress of the north face of Ben Nevis.

Above. The north face of Ben Nevis is largely made of lavas erupted from the volcano that was active around 425 million years ago. There are many similarities between the Ben Nevis and Glencoe volcanoes. Both were explosive in the extreme and had associated pyroclastic and lava flows.

Right. Molten rock of granite composition (shown in red and orange) ascended through the Earth's crust to a position a few kilometres from the surface. In the form of lava it then escaped from the magma chamber in an upwards direction, leaving a void. The overlying rocks then collapsed into that space in a manner defined by a circular fault line as illustrated by the diagram. The surface rocks (shown in blue) that fell into the magma chamber consist of lavas that had been previously erupted and a minor component of sedimentary strata laid down by rivers that flowed across the area. Since then, erosion has cut deep and removed much of the overlying strata. What can be seen today is the fallen block of lava juxtaposed by two generations of granite that filled the magma chamber. The volcano punched a hole through the enclosing Dalradian strata (shown in green) that were already in place.

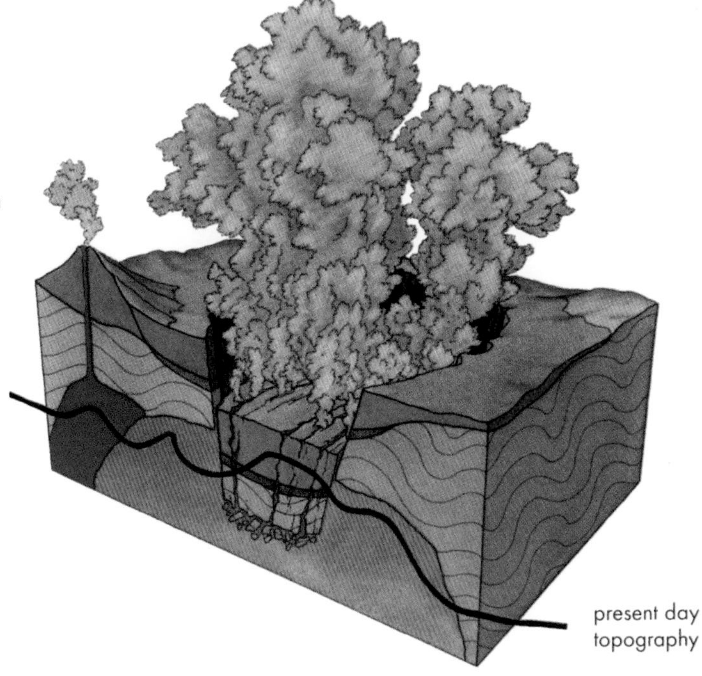

Right. This is a plan view of Ben Nevis today. The great raft of lava that floundered into the magma chamber is the centrepiece to the mountain and is surrounded by the two main pulses of granite that occupied the magma chamber.

Below. Carn Mòr Dearg is made from the granite that formed part of the magma chamber of the Ben Nevis volcano.

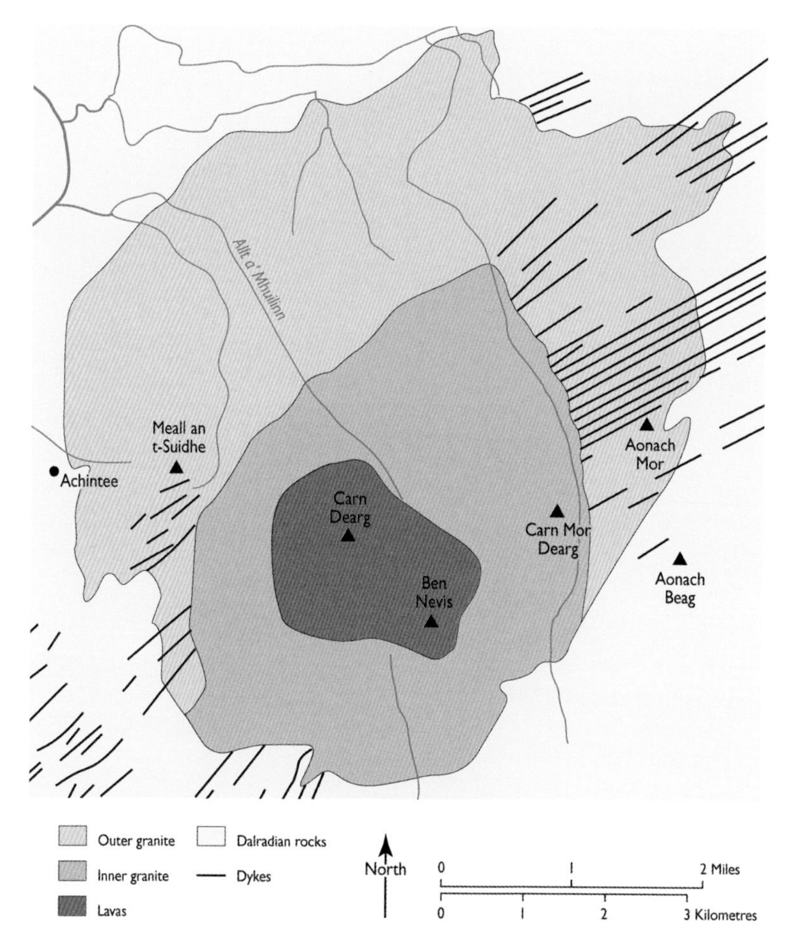

4

A world first at Strontian

The village of Strontian, on the north shore of Loch Sunart, is world famous. It gives its name to strontium, the 38th element of the periodic table, and it is also the place where the mineral strontianite (strontium carbonate) was first discovered. Both claims to fame are a big deal in the world of science. The mineral was first named in 1791 from a specimen recovered from a vein that runs a few kilometres north of the village. And later the new element was isolated and identified from crystals of strontianite by Sir Humphry Davy, who in later life became President of the Royal Society. The mineral was extensively used in the manufacture of cathode ray tubes and early televisions, but when this method of display was superseded by newer technologies its consumption worldwide declined dramatically.

An exquisite specimen of strontianite recovered from near the village that gave this mineral its name.

5
Mind the gap!

The Ben Nevis granite is dated at around 425 million years old and was the last event to leave a significant mark on the landscape of Lochaber and Glencoe until around 58 million years ago, when another series of volcanoes erupted nearby. These more recent lavas emanated from the Mull volcano and engulfed the south-western part of the area, just to the north of the Sound of Mull.

Significant events are recorded elsewhere in Scotland during this hiatus from 425 to 58 million years ago. Coal forests existed during the Carboniferous Period and much of the central belt of Scotland was swathed in tropical rainforests. There is only a tiny patch of rocks from this age remaining, near Rubha an Ridire in the extreme south of Lochaber. Later, when deserts gripped the land, as Scotland continued its northward drift to sit at latitudes similar to that of the Sahara desert of today, the area would have been covered by sand dunes. Sea levels then rose to cover most of Scotland during Jurassic times and dinosaurs roamed the land that we now recognise as the Isle of Skye. In the Cretaceous Period sea levels rose even further to cover most of Scotland. Limited evidence for this major event exists near Lochaline where sands of this age are found and also further north at Beinn Iadain where an extensive sequence of Cretaceous sand, limestone and clays has been recorded. None of these major events left much of a mark on Lochaber and Glencoe. However, the rocks that record these glimpses of past events may have formerly had a wider footprint that has subsequently been reduced during millions of years of erosion by ice, wind and water.

Such yawning gaps in the geological record are the norm, rather than the exception. So it is only by looking at all parts of the country that the geological history of the whole of Scotland can be pieced together.

6
Age of ice

Over two million years ago Scotland, along with much of the northern hemisphere, was plunged into an Ice Age. The climate had previously been warm and temperate, but over a relatively short space of time, temperatures fell on a global scale and ice sheets extended south from the North Pole to submerge the whole of Scotland. This change had a profound effect on the landscapes of the area. Ice sheets and glaciers became established and the flourishing ecosystems from previous times were extinguished as even the highest mountains were swathed in a frozen blanket. Sea levels dropped worldwide in response to the changing climate, as much of the water on the face of the planet was locked up in the growing sheets of ice.

The cold conditions were broken by frequent warmer periods, known as interglacials, but after these short interludes, which lasted around 10,000 years, the ice and snow re-established its icy grip. We are currently enjoying an interglacial period, but glacial conditions may once again be re-established across the country around 55,000 years

This is Greenland today where even the highest mountains are almost submerged under great accumulations of ice and snow. The landscape of the Lochaber and Glencoe area would have looked very similar to this picture for large parts of the Ice Age.

from now. It's a chilling thought, but the cyclic nature of climate change is known from the past and is likely to be a pointer to the future.

To find the reason for temperature changes on a global scale, we need to look beyond our world to the way in which Planet Earth circles the Sun. The orbit around the Sun varies from perfectly circular to elliptical over a period of around 100,000 years. So when we are close to the Sun, temperatures are higher than when the Earth is at the further reaches of its elliptical orbit. There are other irregularities, such as wobbles of the north–south axis around which the Earth spins. These variables lead to regular and predictable changes from glacial to warmer periods that have characterised the Ice Age over the last 2.6 million years.

There have been many advances and retreats of the ice as the climate has changed in response to the variable orbit of the Earth around the Sun. Each new advance of the ice causes yet more erosion and it also has the effect of largely destroying the evidence for what happened during earlier glaciations. The last major advance of the ice took place 22,000 years ago and most of the evidence we can glean about our glacial history comes from these times.

Lairig Eilde in Glencoe. This is known as a U-shaped valley and is a classic example of the landscape changes caused by a glacier grinding its way from higher ground down to an elevation closer to sea level. These moving rivers of ice carry tons of rock and other debris at the interface between the lowest layers of the glacier and the rocks over which the glacier is moving. The glacier sandpapers the landscape, carving out great gouges and smoothing out the contours of the land.

The 'Mer de Glace' near Chamonix in the French Alps is an example of a present-day glacier carving its way through the landscape. This deep valley will in time come to resemble the Lairig Eilde when the climate warms and the ice disappears.

As the ice melted and finally retreated, huge quantities of meltwater were generated. These torrents of water flowed across the landscape, finding the shortest route to the sea. Where progress was impeded by remnant plugs of frozen ice or other obstructions, great lakes of water were trapped behind them. The best evidence for this phenomenon is to be found at Glen Roy near Roy Bridge. Parallel 'benches' were cut into the hillside by waves at the margins of the impounded lake.

As the height of the ice impounding the lake changed, so too did the level of the water, creating a new set of notches in the hillslopes of Glen Roy. This phenomenon was first investigated by eminent scientists from the Victorian era, including Charles Darwin and Louis Agassiz. Darwin thought the notches were cut by the sea, but later accepted they were lake shorelines. Agassiz was a Swiss geologist who saw the work of ice in his native land. He was the first to suggest, when he visited Glen Roy in 1840, that many of the features of the Scottish landscape were also caused by glaciation.

Above. The Parallel Roads of Glen Roy are a series of notches or benches cut into the hillside that run uninterrupted for many kilometres. They are the shoreline marks of a loch that once filled Glen Roy.

Right. This diagram shows the full extent of the parallel roads in the Glen Roy area.

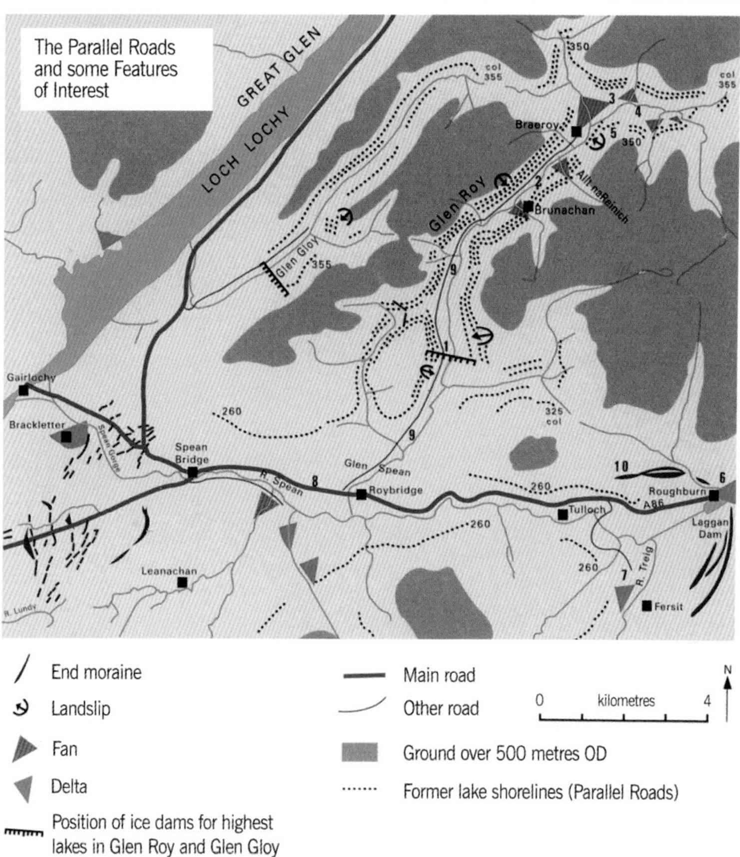

7

The landscape today

Natural landscapes

The varied geology of the area supports a thriving biodiversity. Here are a couple of examples.

Loch Shiel, a deep glacial gouge that runs south-west from Glenfinnan, is one of the biggest oligotrophic (low in nutrients) water bodies in Scotland and supports a different mix of plants and animals from most lochs. A large population of otters have been recorded here. Black-throated and red-throated divers also thrive in this place, although their numbers are largely determined by their breeding success in any given year. The steep-sided banks of the loch are important for their stands of native pinewood and oak and alder woodlands. Nationally rare dragonflies and damselflies have made this place their home, as have rare plants such as bog orchids and Irish lady's-tresses.

Loch Shiel from the north including the Glenfinnan monument.

Left. Black-throated diver.

Right. Otter.

Rannoch Moor stretches eastwards from Buachaille Etive Mòr, the imposing mountain that guards the eastern entrance to Glencoe. It is one of Scotland's biggest and most important areas of natural wetland. It is covered by many designations including Site of Special Scientific Interest, Special Area of Conservation, Special Protection Area, National Nature Reserve and Ramsar site. Fauna included here are: freshwater pearl mussel, Atlantic salmon, Brook lamprey, otters and a large breeding population of black-throated divers. Wet mire, some with quaking surfaces, and blanket bog are the dominant habitats across this internationally important site.

Scottish Natural Heritage's information portal at https://gateway.snh.gov.uk/sitelink is the ideal site to tell you more about the designated sites across the area.

Left. Freshwater pearl mussel.

Below View across Rannoch Moor.

Unstable landscapes

As the ice melted, it left a landscape strewn with a jumble of deposits, known as moraine or glacial till, that had been eroded from the bedrock and carried along by the ice. The sides of the new valleys carved by the ice were left unsupported after the ice melted, and lumps of rock, large and small, tumbled down the slopes. Some rolled downslope as individual blocks, whereas in other places scree slopes were formed with collections of loose debris. Although we may think of the development of landscapes as something that has already happened, changes to familiar views of hills and glens by natural processes continue to this day. If we were able to re-visit these much-loved places thousands of years hence, some would be barely recognisable because so much would have changed in the intervening period. Ours is still a very dynamic Earth.

In Glencoe, a debris cone builds, fed by the crumbling cliffs above.

Quarried landscapes

The Strontian granite provides the raw material for much of the crushed rock aggregate that this part of Scotland produces. Glensanda, an extraction site on the north shore of Loch Linnhe, has been called a superquarry because of its prodigious output of building materials. The stone is not just for domestic consumption; the quarry feeds the voracious and unquenchable international appetite for crushed rock that exists around the world. Since it started work in 1986, it has transported around 6 million tons of granite from the site by ship.

Glensanda superquarry is a large-scale industrial operation set in a rural location, so considerable environmental safeguards are in place to ensure that the surrounding land and sea are not significantly degraded or polluted by the quarrying or shipping operations.

Monstrous landscapes!

Loch Morar is Britain's deepest loch. It is also the subject of an enduring mystery that involves a 10-metre-long serpent called Morag! The legend is perhaps not as well known as that of its more illustrious neighbour in Loch Ness, but the story is just as long-lived. Both lochs were cut

Legend has it that Loch Morar is home to a serpent named Morag!

during glacial times when ice bit deep into the bedrock to excavate these linear loch basins. The ice would have occupied the loch basin from end to end, and from the bed of the loch to the surface of the ice, for thousands of years at a time. So where the monsters were hiding during the 2 million years that the Ice Age lasted is anybody's guess. But we perhaps shouldn't let science get in the way of a good story, particularly one that encourages tourists to come back year after year.

Aesthetic landscapes

The National Scenic Area of Morar, Moidart and Ardnamurchan runs from the most westerly point on the British mainland to include much of the Sound of Arisaig. Part of the citation reads 'The Loch Ailort and Loch nan Uamh are open in aspect, with fine views to the Small Isles. The foreground is a richly wooded shore of rocky promontories, while the waters of the lochs are studded with heather and scrub-covered islets. This richly patterned western prospect typifies for many people the scenery they associate with the romance of "the Road to the Isles".'

Lochaber Geopark

UNESCO European Geopark status is given to areas of Europe that host areas of outstanding geological heritage. Their purpose is to tell the geological story of each designated area through interpretation and educational initiatives. Lochaber Geopark is based in Fort William. Its website, leaflets and on-site interpretation tell the story of the geology and landscapes for residents and visitors alike. The staff also lead guided walks and deliver lectures to the public.

The Sound of Arisaig is part of the Morar, Moidart and Ardnamurchan National Scenic Area.

8
Places to visit

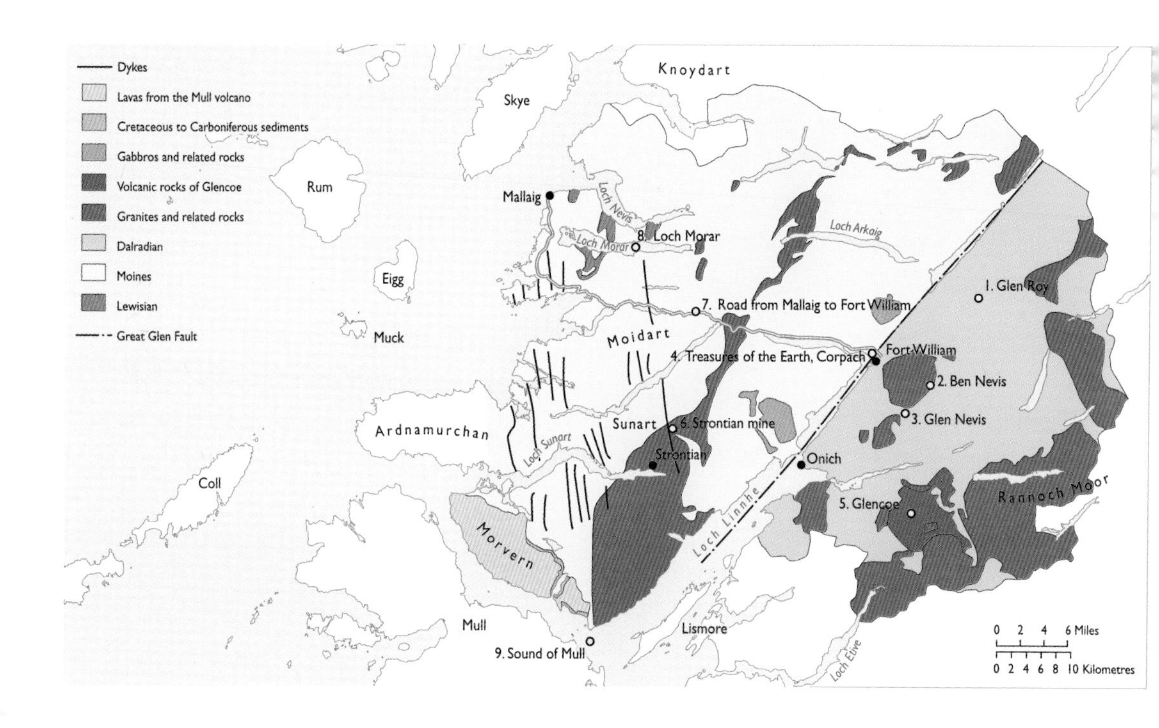

This short section gives a few suggestions of places to visit within the area. Full descriptions of most of these sites are provided earlier in the book. OS Landranger maps 40, 41 and 49 will assist you in navigating between sites.

1. Glen Roy: just to the north of Roy Bridge is the best place to see the 'Parallel Roads', with a viewpoint, car park and interpretation boards at the site. See page 33 and 34 for the site description.

Glen Roy.

2. Ben Nevis: reaching the top of Scotland's highest peak on foot is a slog for only the fittest to attempt. A mountain gondola runs from a point just off the A82 and takes the less energetic visitor to an adjacent peak in the Nevis range. From there, panoramic views of Fort William and the surrounding area are possible. Perhaps wait for a clear day to get the most spectacular view.

Nevis Range gondola – the easy way to the top!

Glen Nevis.

3. Glen Nevis: to the south of Fort William this scenic glen excavated by ice during the recent glaciation provides a gentle stroll beside the River Nevis. The walk ends just a few miles south of Ben Nevis.

4. Treasures of the Earth exhibition: this is located on the Road to the Isles on the outskirts of Fort William. It immerses the visitor in a world of rare crystals and fossils. An entrance fee is charged. There is good parking.

5. Glencoe: a must for any visitor to the area. There is a National Trust for Scotland visitor centre at the western end of the glen and many parking places to stop and photograph the stunning scenery along the length of the glen.

Glencoe.

6. Strontian mine: the source of the mineral strontianite is located 3 kilometres north of the village of Strontian. The main workings are dangerous, but looking over the protective railing you will get a good view of the old opencast and mine workings. There are also spoil tips that can be visited in safety on the other side of the road.

Strontian mine.

Road cutting near Lochailort.

7. Road from Mallaig to Fort William: this provides an excellent 'section' through the Moine geology, with frequent road cuttings that show the folded Moine rocks. The illlustration on page 18 is a summary of what geologists have pieced together by studying these exposures and also those to the north and east in Moidart and Sunart.

Loch Morar.

8. Loch Morar: this peaceful place is easily explored along the minor road running eastwards from Morar village for a short distance on the north shore of the loch. No sightings of Morag the monster are guaranteed, but it's a route that shows off how stunning Scotland's land and waterscapes can be.

9. Boat trip along the Sound of Mull: this is a good way to see the varied geology of the area. Follow the changing rock types on the geological map on page 10 from granite at the eastern end of the Sound, through the Moines and finally the parallel layers of lava that flowed from the Mull volcano.

North shore of the Sound of Mull.

Acknowledgements and picture credits

Thanks are due to Professor Stuart Monro OBE FRSE and Moira McKirdy MBE for their comments and suggestions on the various drafts of this book. I also thank Debs Warner, Mairi Sutherland, Andrew Simmons and Hugh Andrew from Birlinn for their support and direction. Mark Blackadder's book design is up to his usual high standard. Scottish Natural Heritage, in association with the British Geological Survey, published the *Landscape Fashioned by Geology* series that was the precursor to the new *Landscapes in Stone* titles. I thank them both for their permission to use some of the original artwork and photography in this book. Kathryn Goodenough and David Stephenson wrote the original text for *Ben Nevis and Glencoe – A Landscape Fashioned by Geology*, and Douglas Peacock, Frank May and John Gordon wrote the original text for *Glen Roy – A Landscape Fashioned by Geology*. Both books influenced aspects of this publication. I dedicate this book to David Stephenson who has done as much as anyone, past or present, to unravel the geology of the Scottish Highlands. David worked for the British Geological Survey for over three decades and has published a prodigious number of books, articles and scientific papers on various aspects of Scottish geology.

Picture credits

ii–iii Banana Pancake/Alamy Stock Photo; 6 Lorne Gill; 10 drawn by Jim Lewis; 11 Harry Feather; 12 North Wind Picture Archives/Alamy Stock Photo; 13 (upper) drawn by Robert Nelmes, (lower) drawn by Jim Lewis; 14 (upper) drawn by Robert Nelmes, (lower) drawn by Jim Lewis; 15 Andrew Hagen; 16 (left) British Geological Survey © NERC, (right) Alfred Pasieka/Science Photo Library; 17 (upper) Alan McKirdy, (lower) John A. Cameron; 18 drawn by Jim Lewis; 19 (upper) drawn by Jim Lewis; (lower) Lorne Gill; 20 (upper) Robert Nelmes/SNH, (lower) drawn by Craig Ellery; 21 Lorne Gill/SNH; 22 drawn by Jim Lewis; 23 Lorne Gill; 24 (left) CPC Collection/ Alamy Stock Photo, (right) Peter Kokelaar/BGS; 25 Kathryn Goodenough/BGS; 26 (top, left and right) Peter Kokelaar/BGS, (lower) Noel Williams; 27 (upper) Alex Gillespie Photography, (lower) Claire Hewitt; 28 (upper) drawn by Jim Lewis, (lower) Noel Williams; 29 © National Museums Scotland; 31 John Gordon; 32 Patricia and Angus Macdonald/Aerographica/SNH; 33 Moira McKirdy; 34 (upper) Mark Godden, (lower) Craig Ellery; 35 David Gonzalez Rebollo; 36 (left) Erni, (right) Michal Ninger; 37 (upper) Antonio Davila Santos, (lower) robertharding/ Alamy Stock Photo; 38 Lorne Gill; 39: Glensanda Superquarry © Michael Jagger (cc-by-sa/2.0); 40 Mike Kipling Photography/Alamy Stock Photo; 41 John A. Cameron; 42 drawn by Jim Lewis; 43 (upper) Lorne Gill, (lower) David McElroy; 44 Luis Abrantes; 45 (upper) Nitsawan Katerattanakul, (lower) Moira McKirdy; 46 (upper), Alan McKirdy; 46–7 Alan McKirdy; 47 Moira McKirdy